Table of contents

1 Ways of thinking in mathematics you have learned

In mathematics, it is not only important to do calculations and explore figures. The "ways of thinking" that you found while learning mathematics are useful tools not only in mathematics but also to solve problems in your surroundings.

Until now in mathematics, the focus has been to find the following nine "ways of thinking." These "ways of thinking" can be used not only for studying in junior high school, but also in various situations when you grow up.

1 Setting the unit.

Unit

Once you set the unit, you were able to represent a size by how many times of the unit using a number.

2 If you try to arrange...

Align

When comparing sizes and performing various calculations, you were able to think by aligning the number place and the unit.

Also, you were able to compare speed by aligning the value based on the unit of time or distance.

3 If you try to separate...

Separate

You were able to solve a complete problem that you couldn't understand as it was presented, by separating it into smaller sections.

4 If you try to summarize...

Summarize

After giving the answer, by summarizing problems and thoughts from people in similar situations, you found out that even when the way to express was different, you were able to find cases with the same matter.

5 If you represent in other way...

Other way

You were able to understand and represent a problem in an easier way by replacing the problem with a math expression, diagram, table, or graph.

6 If you try to change the number or figure...

Change

By changing part of the problem, you were able to discover new problems or better understand the problem.

7 Can you do the same or similar way?

Looks same

Thinking about using the same method you've done so far, you were able to find a clue for a new problem.

8 Is there a rule?

Rule

By examining some cases regarding certain things, you were able to find out there was a rule.

9 You wonder why?

Why

When you explained to others the reason why, it was easier to understand by clarifying the reasons and being able to explain in order.

These "ways of thinking" can be used individually or in combination depending on the case.

1 Setting the unit.

 1

Let's explore the number 3758.

① What does the "3" represent?

Daiki

It's the number in the thousands place. It means 3000.

It means three units of 1000. If we think of 1000 as one unit, it represents 3 times of that.

Nanami

② Let's use Nanami's idea and represent 3758 in a math sentence.

Nanami's idea

Since there are 3 units of 1000, this represents [ ] × 3

The numbers on the other places can be [ ] × 7

considered in the same way: [ ] × 5

[ ] × 8

3758 is the sum of the results above, so it can be represented as follows:

3758 = [ ] × 3 + [ ] × 7 + [ ] × [ ] + [ ] × [ ]

If you use the way of thinking "Setting the unit," you can consider numbers in each place such as 1000 or 100 as "one unit," and represent how many units there are in a math expression.

 1 Let's explore the decimal number 31.5.

① If 0.1 is considered as one unit, how can you represent the number 31.5?

② Let's represent what you explained on ① in a math sentence.

[Length]

2 As for the straight lines drawn on the blackboard, which one is longer? Can you compare them without using a ruler?

We can compare them by looking at how many erasers are equivalent to the length of the lines.

Yui

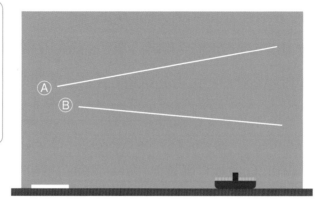

We can also compare them using a chalk.

Hiroto

① How many pieces of chalk are equivalent to the length of Ⓐ? Let's explore it by using a compass.

② How many pieces of chalk are equivalent to the length of Ⓑ? Let's explore it by using a compass.

③ Which is longer, Ⓐ or Ⓑ?

Regarding length, you can represent it with numbers such as "equal to 5 pencils" or "equal to 12 erasers." In these cases, one pencil or one eraser is considered as "one unit" and the length is represented by the number of units. But, if all the people in the world represent length by their own "one unit," then lengths cannot be compared.

First of all, as a common "one unit," the length of 1 m was decided. Next, $\frac{1}{100}$ of 1 m was decided as 1 cm and $\frac{1}{10}$ of 1 cm was decided as 1mm. In this way, "one unit of length" was made common in the world and we measure by these units.

[Area and weight]

3   **Let's explore a square with a side length of 1 m 30 cm.**

①   Considering the area of a square with a side length of 1 cm as one unit, how many units are there in the original square?

②   Considering the area of a square with a side length of 10 cm as one unit, how many units are there in the original square?

As with lengths, areas can also be represented with numbers when "one unit of area" is decided. "One unit of area" was defined with units such as 1 cm² or 1 m², and the areas represented by a number of units are simply called "area."

4   **The weight of a pair of scissors and two batteries is explored by using a balance. When one-yen coins are used, the objects are balanced as shown below.**

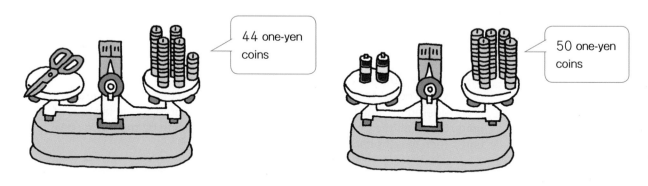

44 one-yen coins

50 one-yen coins

①   Which is heavier, "a pair of scissors" or "two batteries"?

②   Each one-yen coin weighs 1 g. How many grams is the weight of a pair of scissors and one battery respectively?

Also for weight, various weights can be represented by using the unit "gram" and define "one unit of weight" as 1 g.

2 If you try to arrange...

[Calculations of whole numbers and decimal numbers in vertical form]

1 Let's solve the following calculations in vertical form.

① 3250 + 466　② 13.4 + 2.57　③ 5130 − 285　④ 24.1 − 0.65

When you add or subtract whole numbers and decimal numbers in vertical form, you align the digits of the numbers according to their places.

[Comparison of the size of fractions]

2 Which of the fractions shown on the right is larger?　$\frac{2}{3}$　$\frac{5}{7}$

I can change them into decimal numbers and then compare them.

Daiki

I can align the denominator and then compare them.

Nanami

① Let's compare the sizes by using Daiki's idea.

Daiki's idea

If $\frac{2}{3}$ = ☐ ÷ ☐ is found by rounding off the thousandths place, it's about ☐.

If $\frac{5}{7}$ = ☐ ÷ ☐ is found by rounding off the thousandths place, it's about ☐.

Therefore, using an inequality sign, it can be represented as $\frac{2}{3}$ ☐ $\frac{5}{7}$.

② Let's compare the sizes by using Nanami's idea.

Nanami's idea

If the fractions were $\frac{2}{7}$ and $\frac{5}{7}$, the sizes could be compared.

Because $\frac{2}{7}$ is ☐ sets of $\frac{1}{7}$ and $\frac{5}{7}$ is ☐ sets of $\frac{1}{7}$, $\frac{5}{7}$ is ☐ than $\frac{2}{7}$.

So, without changing the ☐ , the denominators of $\frac{2}{3}$ and $\frac{5}{7}$ can be aligned.

If I change to a common denominator and compare, $\frac{2}{3}$ = ☐ , $\frac{5}{7}$ = ☐

Therefore, ☐ is larger.

③ Yui said that she can also compare sizes by aligning the numerator.

Let's explain how she would compare.

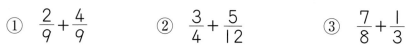

When you compare the size of fractions, if you "align" the denominator or numerator, the following properties of fractions can be used:
With the same denominator, the fraction with a larger numerator is larger.
With the same numerator, the fraction with a smaller denominator is larger.

1 Let's solve the following calculations.

① $\frac{2}{9} + \frac{4}{9}$

② $\frac{3}{4} + \frac{5}{12}$

③ $\frac{7}{8} + \frac{1}{3}$

④ $1\frac{1}{2} + \frac{4}{5}$

⑤ $2\frac{1}{6} + 1\frac{3}{10}$

⑥ $2\frac{8}{9} + \frac{2}{3}$

⑦ $\frac{6}{7} - \frac{1}{7}$

⑧ $\frac{5}{8} - \frac{1}{4}$

⑨ $1 - \frac{2}{5}$

⑩ $1\frac{2}{11} - \frac{1}{3}$

⑪ $2\frac{7}{8} - 2\frac{1}{12}$

⑫ $3\frac{1}{2} - 1\frac{5}{6}$

[Comparison of speed]

When comparing speed, the way of thinking "aligning" is important.

3 **Let's compare the speed of the following two trains.**

┌─**Train A**─────────────────┐
It runs a distance of 15 km, between

stations C and D, in 8 minutes.

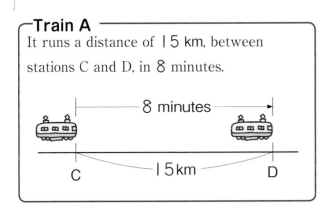

┌─**Train B**─────────────────┐
It runs a distance of 12 km, between

stations E and F, in 7 minutes.

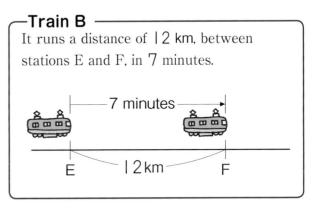

Nanami

In order to determine the fastest person in a race, both people should run the same distance and then compare the speed by the time to reach the goal.

There are also races in which the person who runs the longest distance in a fixed period of time becomes the winner.

Hiroto

① Let's compare the speed of Train A and Train B by aligning the distance.

Align the distance by using the least common multiple of 15 km and 12 km.

15 × 4 = ☐ (km)

12 × ☐ = 60 (km)

The duration which Train A takes to travel 60 km is 8 × ☐ = 32 (min).

The duration which Train B takes to travel 60 km is 7 × 5 = ☐ (min).

Therefore, Train ☐ is faster, as it takes less time to travel the same

distance.

9

② Let's compare the speed of Train A and Train B by aligning the time.

> Align the time by using the least common multiple of 8 minutes and 7 minutes.

Let's continue with the following steps and write in your notebook in the same way as in ①.

③ By using the way of thinking "setting the unit," you can find how many kilometers can be traveled in 1 minute and how many minutes it takes to travel 1 km.

For example, if you consider the distance that Train A travels in 1 minute as x km, then you can find x as follows:

Speed	Distance
x km	15 km
1 min	8 min

Time

$$x \times 8 = 15$$
$$x = 15 \div 8$$
$$x = 1.875$$

In the same way, by considering the distance that Train B travels in 1 minute as x km, let's find x and compare the speed of Train A and Train B.

> The speed of moving things can be compared by using the way of thinking "aligning" as follows:
> Comparing time by aligning the moving distance.
> Comparing the moving distance by aligning time.

3 If you try to separate...

[Multiplication]

Let's think about how to calculate 23×6.

① Daiki explained how to calculate 23×6 as follows. Let's write the number that applies in each ☐ .

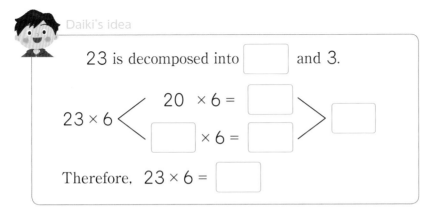

Daiki's idea

23 is decomposed into ☐ and 3.

$$23 \times 6 \begin{cases} 20 \times 6 = \boxed{} \\ \boxed{} \times 6 = \boxed{} \end{cases} \boxed{}$$

Therefore, $23 \times 6 = \boxed{}$

② By using Daiki's idea as a reference, let's write down how to calculate 34×7 in your notebook.

In order to use what you have learned before, the way of thinking "decomposing" enables you to multiply large numbers based on the multiplication table.

[Even numbers and odd numbers]

Let's categorize the following whole numbers as even or odd numbers.

35   101   52   4   2020   697   800   943

Based on the remainder when the whole numbers are divided by 2, they can be categorized.

3

Let's classify the following quadrilaterals.

Ⓐ  Ⓑ 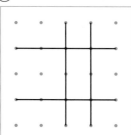 Ⓒ 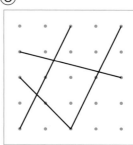 Ⓓ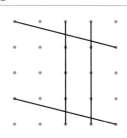

Ⓔ 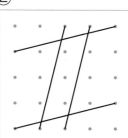 Ⓕ 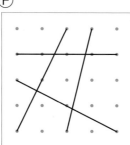 Ⓖ 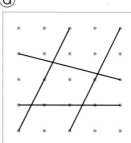 Ⓗ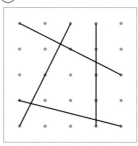

① Let's choose quadrilaterals that have only one pair of parallel sides.

② Let's choose quadrilaterals that have two pairs of parallel sides.

③ Let's choose quadrilaterals that do not have a pair of parallel sides.

Square: quadrilateral in which all 4 angles are right angles and all 4 sides have the same length.

Rectangle: quadrilateral in which all 4 angles are right angles.

Trapezoid: quadrilateral in which one pair of opposite sides are parallel.

Parallelogram: quadrilateral in which two pairs of opposite sides are parallel.

Rhombus: quadrilateral in which all 4 sides have the same length.

 Based on the angles and sides, quadrilaterals can be classified.

4 If you try to summarize...

[Large numbers]

1 **Let's think about the structure of large numbers.**

① What is the number that gathers 10 sets of 1000?

② What is the number that gathers 10 sets of 10 million?

③ Let's explain about 600000. Write the number that applies in each of the

following ☐ .

600000 is a number that gathers ☐ sets of 100000.

Also, it can be said that gathers ☐ sets of 10000 or ☐ sets of 1000.

If you gather 10 sets of 100, it becomes 1000. So, if you put together by 10, you move to the next value place. It can be said that putting together by 10 is "doing by 10 times."

[Table]

2 **Only one favorite subject of students from 4th Grade to 6th Grade was surveyed and summarized in a table. Let's think about the table.**

4th graders' favorite subjects

Subject	Number of students
Japanese	26
Mathematics	18
Science	22
Social studies	16
Other	32
Total	114

5th graders' favorite subjects

Subject	Number of students
Japanese	28
Mathematics	22
Social studies	23
Social studies	24
Other	18
Total	115

6th graders' favorite subjects

Subject	Number of students
Japanese	25
Mathematics	24
Science	26
Social studies	22
Other	16
Total	113

① To compare in an easier way, the data was summarized in a single table. Let's fill in the blanks to complete the table.

Favorite Subjects   (children)

subjects \ students	4th graders	5th graders	6th graders	Total
Japanese		28	25	
Mathematics	18			
Science		23	26	71
Social studies	16			62
Other		18		
Total	114		113	342

② In the above table, what do the following numerals represent?

   ⓐ 25 ⓑ 62 ⓒ 342

③ Which subject has the largest total number of students?

[Figures]

3 The properties of regular polygons were summarized. Let's fill in the blanks in the table. However, for "line symmetric figure" and "point symmetric figure," write ◯ if applies, otherwise write ✕.

	Equilateral triangle	Square	Regular pentagon	Regular hexagon	Regular octagon	Regular dodecagon
Number of sides						
Number of vertices						
Size of one angle						
Line symmetric figure						
Number of lines of symmetry						
Point symmetric figure						

It is easy to understand if the properties of the figures are organized in a table.

[Meaning of multiplication]

4 **Let's think about the problems considered by the following three children.**

Daiki's problem

There is a metal bar that is 1 m long and weighs 9 kg. How many kilograms is the weight of 5 m of this metal bar?

Nanami's problem

There is a string that is 9 m long. How many meters is the length of the string that is 5 times of the original?

Hiroto's problem

How many square centimeters is the area of a rectangle with a length of 9 cm and a width of 5 cm?

① As for the problems, which of the following is true for each of them.

Let's choose one from Ⓐ～Ⓒ.

    Ⓐ   A problem that finds the area.

    Ⓑ   A problem that uses a measure per unit quantity.

    Ⓒ   A problem that uses a ratio.

② Let's represent each problem in a math expression and find the answer.

Math expressions can integrate various solutions and meanings.

5 If you represent in other way...

[Table and graph]

The table on the right shows the result of a survey on whether a dog or cat is kept at home in a certain class. Let's think about this.

Have only a dog	9 children
Have only a cat	6 children
Have both a dog and a cat	3 children
Do not have any dogs or cats	18 children

① How many children are there in the class?

② Let's represent the ratio of children who have a dog to the total number of children as a fraction.

③ What percentage of children have a cat?

④ When represented in a circle graph, how many degrees does each child represent?

⑤ Let's represent the number of children who only have a dog in the following circle graph. Let's represent the ratio with an angle.

⑥ Let's complete the following circle graph.

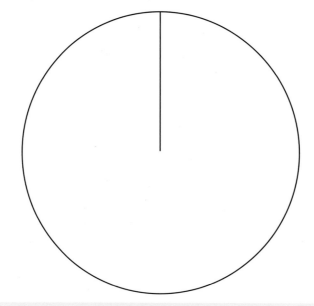

If you represent numbers in different ways, such as ratios or graphs, you can identify relationships at a glance and compare with other data.

[Combination]

2 Teams A, B, C, D, and E will play baseball games. If each team plays one game with all other teams, how many games will be played in total?

① Let's think by using the methods of the following children.

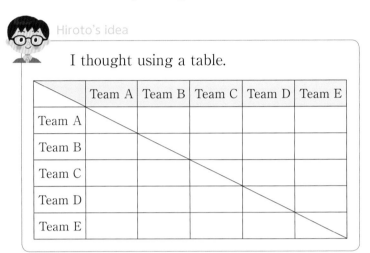

Hiroto's idea

I thought using a table.

	Team A	Team B	Team C	Team D	Team E
Team A					
Team B					
Team C					
Team D					
Team E					

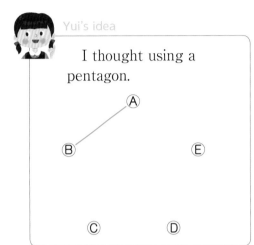

Yui's idea

I thought using a pentagon.

② In order to understand the outcome at the same time, which idea is more convenient, Hiroto's idea or Yui's idea?

[Mathematical letters and sentences]

3 There is a quadrangular prism with an area of the base of 12 cm². Let's represent the relationship between the height and the volume in a math sentence.

① Let's consider the height as x cm and write a math sentence to find the volume of the quadrangular prism.

② When the volume is 192 cm³, how many centimeters is the height?

There is also a way of thinking in which unknown numbers are replaced by mathematical letters.

1 Let's find the number that applies to x.

① $14 + x = 34$　② $x + 8 = 42$

③ $x - 16 = 9$　④ $x \times 7 = 63$

17

6 If you try to change the number or figure...

[Ways of finding the area of a trapezoid]

1

Let's find the area of the trapezoid shown on the right in various ways.

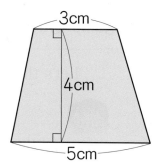

① Let's write the words and numbers that apply in the following ☐.

Daiki's idea

I ☐ the trapezoid into two triangles.

The area of triangle Ⓐ is ☐ × 4 ÷ 2 = ☐

The area of triangle Ⓑ is 5 × ☐ ÷ 2 = ☐

Since the area of the trapezoid is found by

the area of ☐ + the area of ☐ , then

☐ + ☐ = 16. Therefore, the area of the trapezoid is 16 cm².

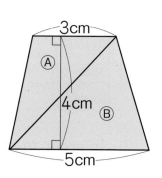

② Let's explain the ways of thinking of the following children.

Hiroto's idea

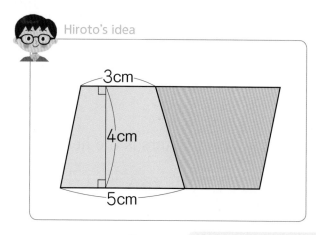

Yui's idea

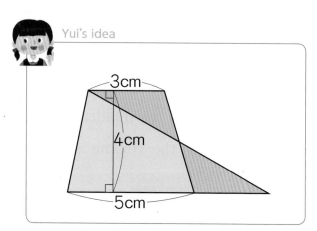

When finding the area of a figure, you can change the figure to a previously learned figure and use the corresponding method to calculate the area.

[Various quadrilaterals]

2 **Let's draw various parallelograms.**

① Let's draw a parallelogram in which the length of side AB is 4 cm and the length of side BC is 6 cm.

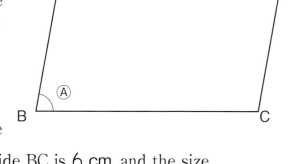

② Let's draw a parallelogram in which the length of side AB is 6 cm and the length of side BC is 6 cm. Also, which kind of quadrilateral is it?

③ Let's draw a parallelogram in which the length of side AB is 4 cm, the length of side BC is 6 cm, and the size of angle Ⓐ is 90°. Also, which kind of quadrilateral is it?

If you make a small change in the conditions of the parallelogram, it may become a rhombus or a rectangle.

[Creation of math problems]

3 On the following problem, let's fill in one of the ☐ with an x and in the rest of the ☐ with numbers, and create various math problems.

☐ m of wire weigh ☐ g. When there are ☐ m of wire, the weight is ☐ g. Let's find the number that applies to x.

If the conditions of a problem are changed depending on what you want to find, then the math sentence also changes and you can create various problems.

19

7 Can you do the same or similar way?

[Multiplication and division of decimal numbers in vertical form]

1 Let's think about how to calculate 3.08 × 7.5.

```
      3. 0 8  ─────────  [   ] times  ──────▶       3 0 8
  ×     7. 5  ─────────  [   ] times  ──────▶   ×     7 5
      1 5 4 0                                     1 5 4 0
    2 1 5 6                                     2 1 5 6
  2 3. 1 0 0  ◀────────  [   ]  ────────────  2 3 1 0 0
```

The case decimal number × decimal number in vertical form is calculated in the same way as the calculation with whole numbers, assuming that there is no decimal point. The number of digits after the decimal point of the product is the same as the sum of those of the decimal numbers.

1 Let's solve the following calculations in vertical form. However, for exercise ⑧, round off the thousandths place and find the nearest hundredths place round number.

① 1.4 × 97　　② 0.8 × 5.3　　③ 2.76 × 3.5　　④ 8.2 × 1.09

⑤ 72 ÷ 1.6　　⑥ 0.4 ÷ 0.25　　⑦ 3.12 ÷ 6.5　　⑧ 3.7 ÷ 2.1

[Multiplication and division of fractions]

2 Let's solve the following calculations.

① $\frac{3}{8} \times 6$　　② $1\frac{8}{10} \times 5$　　③ $\frac{2}{7} \times \frac{1}{4}$　　④ $\frac{4}{9} \times 2\frac{7}{10}$

⑤ $\frac{5}{6} \div 10$　　⑥ $\frac{3}{12} \div \frac{2}{3}$　　⑦ $3\frac{1}{2} \div \frac{7}{8}$　　⑧ $2\frac{1}{5} \div 1\frac{4}{11}$

3 **Let's think about how to draw triangle DEF that is a 2 times enlarged drawing of triangle ABC.**

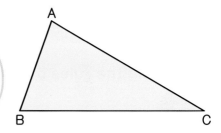

① The drawings of three children are shown below. Let's explain how each child draw the triangle.

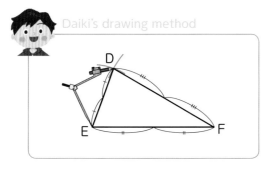

Daiki's drawing method

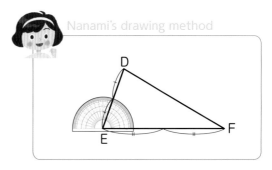

Nanami's drawing method

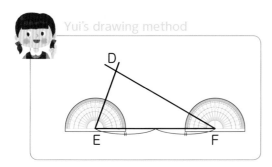

Yui's drawing method

Enlarged and reduced drawings can be drawn with the same methods for drawing congruent figures.

② Straight line EF was drawn with a length that is 2 times the length of side BC. Let's draw triangle DEF using the position of vertex D, which is the corresponding angle to angle A, referred to the drawing methods shown above.

E ——————————————————————————— F

8 Is there a rule?

[Finding the rules of figures]

 Let's answer about the following triangles.

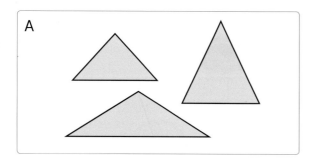

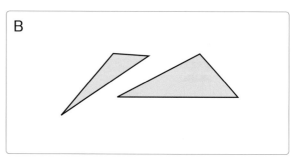

① Let's find common properties in the triangles in group A.

② From the properties found in ①, let's find the properties that also apply to the triangles in group B.

Let's try to explore angles and sides.

2 **Let's find common properties in the following quadrilaterals.**

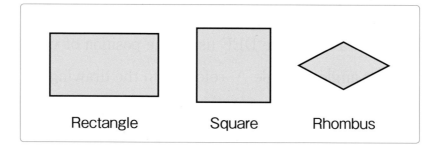

Rectangle   Square   Rhombus

By "finding things that do not change" in various figures, you can find common properties of these figures.

[Two quantities changing together]

3 **Let's look at the table below and answer the following questions.**

Length and width of a rectangle

Length (cm)	1	2	3	4	5	6	7	8	9
Width (cm)	19	18	17	16	15	14	13	12	11

① What kind of rectangle can a rectangle that changes as shown in the table be called? Let's think to find the number that doesn't change.

② Consider the length as x cm and the width as y cm. Based on the rule found in ①, let's represent y with a math sentence that uses x.

③ Does it ever become a square? How many centimeters is the length of the side at that time?

4 **A water tank contains water to a depth of 5 cm. The table below represents the changes that occur when more water is poured in. Let's answer the following questions.**

Time and depth when water is poured in

Time (min)	0	1	2	3	4	5	6	7	8
Depth (cm)	5	6	7	8	9	10	11	12	13

① As for the relationship between time and depth, what doesn't change?

② Consider time as x minutes and depth as y cm. Based on the rule found in ①, let's represent y with a math sentence that uses x.

③ The depth of a water tank is 50 cm. How many minutes will it take to fill the tank?

By finding rules, you can predict what is not written in the table.

9 You wonder why?

[Triangles]

1 Consider A as the center of a circle with a radius of 4 cm and points B and C on the circumference, and draw triangle ABC. Let's think about triangle ABC.

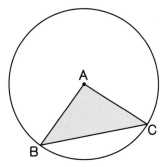

① What kind of triangle is triangle ABC when side BC has a length of 6 cm? Let's also explain the reason.

② How many degrees should be the size of angle A such that triangle ABC becomes an equilateral triangle? Let's also explain the reason.

Hiroto

When explaining the reasons, if you use words like "first," "next," "then," and "therefore," it becomes easier to understand.

An isosceles triangle has two sides with equal length and two angles with equal size.

Also, an equilateral triangle has all three sides with equal length and all three angles with equal size.

To describe an isosceles triangle, you need to explain that there are two sides with the same length. To describe an equilateral triangle, you need to explain that all three sides have the same length.

2 From the following triangles, let's find congruent triangles.

Also, let's explain why you can say that.

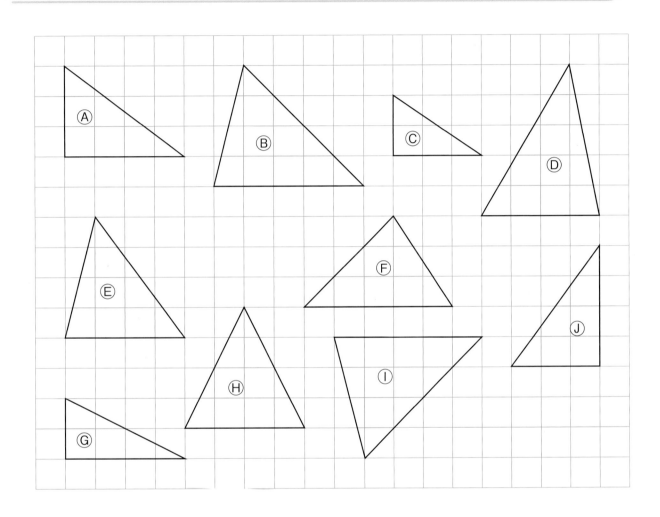

In congruent figures, the corresponding sides have equal length and the corresponding angles have equal size. By focusing on that, you can find congruent triangles.

2 Bridge to the Junior High School

What is a negative number?

Want to know

1 One day, the temperature in Sapporo was negative 4℃. Let's look at the thermometer on the right and discuss how many degrees negative 4℃ is lower than 0℃.

How much is one thermometer scale?

Hiroto

Every scale has the same width.

Nanami

Purpose Can a number with a "minus" be considered in the same way as a number greater than 0?

A temperature that is 4℃ lower than 0℃ is written as −4℃ using the "−" and read "negative four degrees Celsius." "Negative 4" means the number which is 4 less than 0. By using 0 as a origin, we use "−" to express numbers that are less than 0 and we sometimes use "+" to express numbers that are greater than 0.

A number that is greater than 0 is called a positive number and a number that is less than 0 is called a negative number.

0, positive numbers, and negative numbers can be represented on a number line as well as other numbers we have learned so far.

Way to see and think

You can see the thermometer as the number line.

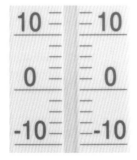

Replace thermometer with a straight line.

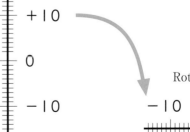

Rotate the line.

If you draw a number line including negative numbers, 0 no longer represents nothing but it is considered as a number. It identifies the origin to separate positive and negative numbers.

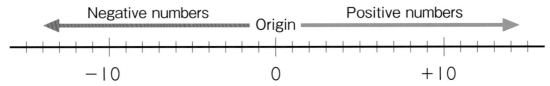

On a number line, we call the direction to the right as "positive direction" and the direction to the left as "negative direction." The numbers get larger as you go to the right, and they get smaller as you go to the left.

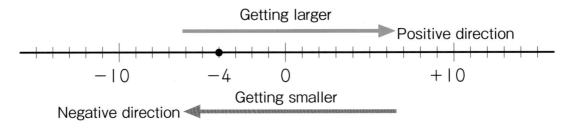

When − 4 is represented in the number line, its position is indicated by the above symbol ●.

Want to confirm

1 Let's represent the following numbers in the number line below.

Ⓐ − 5   Ⓑ + 3   Ⓒ − 1   Ⓓ + 8

Want to try

2 By using the above number line, let's think about the difference between − 4 and 5.

Way to see and think

Whether it's a positive number or a negative number, the size of one scale remains the same.

Summary

A number with a minus sign (−) is a negative number, indicating that the number is less than 0. Also, you can find the difference between positive and negative numbers by using a number line.

In Junior High School, you will learn addition and subtraction using negative numbers.

Card Game

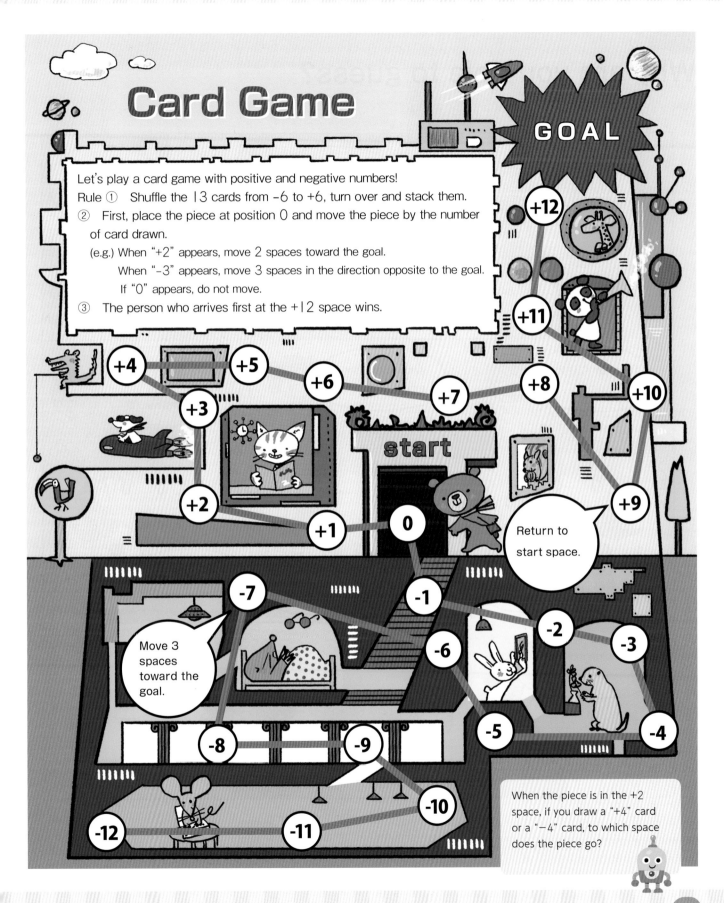

Let's play a card game with positive and negative numbers!

Rule ① Shuffle the 13 cards from −6 to +6, turn over and stack them.

② First, place the piece at position 0 and move the piece by the number of card drawn.

(e.g.) When "+2" appears, move 2 spaces toward the goal.

When "−3" appears, move 3 spaces in the direction opposite to the goal.

If "0" appears, do not move.

③ The person who arrives first at the +12 space wins.

GOAL

Return to start space.

Move 3 spaces toward the goal.

start

When the piece is in the +2 space, if you draw a "+4" card or a "−4" card, to which space does the piece go?

Why are you able to guess?

Problem Why was Daiki able to guess the birth month?

30

Let's think about the quiz shown on the previous page using your birth month.

As my birth month is June,
$6 \times 5 = 30$
$30 + 20 = 50$
$50 \times 2 = 100$
$100 - 40 = 60$

Nanami

If the birth month is June, the result is 60. If the birth month is December, the result is 120. So ...

Hiroto

Purpose Can we confirm why the birth month can be guessed?

Way to see and think

① Let's examine all numbers from January to December.

Do you know what kind of rules are there?

② In this quiz, the answer is 10 times the birth month. Why is it 10 times? Let's compare between the cases of February and x month.

	Case: February	Case: x month
(1) Multiply the number of the birth month by 5.	→ 2×5	$x \times 5$
(2) Add 20 to the result.	→ $2 \times 5 + 20$	$x \times 5 + 20$
(3) Multiply the result by 2.	→ $(2 \times 5 + 20) \times 2$	$(x \times 5 + 20) \times 2$
(4) Subtract 40 from the result.	→ $(2 \times 5 + 20) \times 2 - 40$	$(x \times 5 + 20) \times 2 - 40$

Want to explain

③ Hiroto considered a math sentence using x as birth month in ② as follows. Let's explain Hiroto's idea.

Hiroto's idea

$$(x \times 5 + 20) \times 2 - 40 = x \times 5 \times 2 + 20 \times 2 - 40$$
$$= x \times 10 + 40 - 40$$
$$= x \times 10$$

In Junior High School, you will learn operations with math sentences using mathematical letters.

Summary

The use of mathematical letters can make the calculation mechanism easier to understand.

2

The following objects are in balance on the scale:

2 batteries of the same type and 3 one-yen coins on the left pan

25 one-yen coins on the right pan

1 one-yen coin weighs 1 g.
How many grams does one battery weigh?

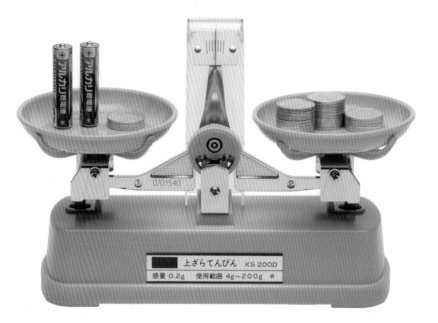

Daiki

The fact that they are balanced means that the weight on the left pan is the same as the weight on the right pan.

If we represent the weight of the battery with math letters ...

Yui

Purpose How can we find an unknown number?

① If x g represents the weight of the battery, let's express the weight of the objects on the left and right pan.

left pan ⬚ g

right pan ⬚ g

You could represent a number in the form of a math expression.

② Let's represent in a math sentence that the left and right pans are balanced.

③ What should we do on the right pan to keep it balanced even if we take ㅣ one-yen coin away from the left pan?

④ The objects are still in balance after the following operation. What was done? What kind of operation is performed in the math sentence shown at the lower right?

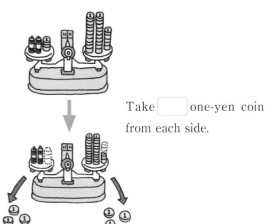

Take ☐ one-yen coin from each side.

If the weight of a battery is represented as x g in a math sentence,

$$x \times 2 + 3 = 25$$

Subtract ☐ from both sides.

$$x \times 2 = 22$$

⑤ After performing the operation in ④, another operation was performed as shown below and the scale remained balanced. What operation was done?

Way to see and think
The left and right sides of the equal sign are the same quantity.

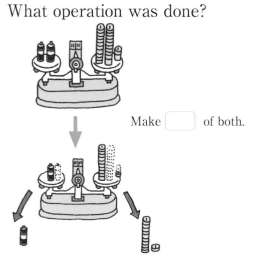

Make ☐ of both.

If the weight of a battery is represented as x g in a math sentence,

$$x \times 2 = 22$$

Divide both sides by ☐.

$$x = 11$$

Summary

As with the scale, we can find unknown numbers in a math sentence by performing the same operation on the left and right sides of the equal sign.

In Junior High School, you will learn how to find the numbers that apply to mathematical letters using various math sentences.

Can you draw only with a ruler and a compass?

 Problem   Can we draw perpendicular and parallel straight lines only with a ruler and a compass?

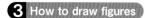

Want to know   How to draw perpendicular straight lines

1

Nanami used a ruler and a compass to draw a straight line that is perpendicular to straight line ⓐ and passes through point A as shown on the right. Let's explain how she drew it.

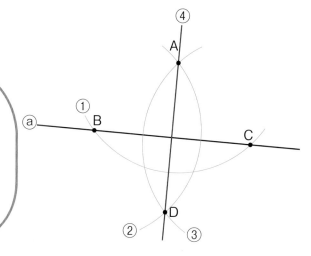

 Nanami: If I connect points in the order of A, B, D, C, and A...

Why can it be said that it is perpendicular? :Hiroto

🌱 Purpose   Can we draw perpendicular and parallel straight lines only with a ruler and a compass?

 Way to see and think

Why can you draw a perpendicular straight line when you draw points B, C, and D?

①   What kind of figure becomes after connecting points in the order of A, B, D, C, and A?

②   As for the diagonals of a rhombus, what kind of properties do they have?

Want to confirm

1 In the following diagram, let's draw a straight line that is perpendicular to straight line ⓐ and passes through point A by using a ruler and a compass. Also, let's confirm that the straight line is perpendicular to straight line ⓐ.

• A

ⓐ   B

Want to try

2 In the above diagram, let's draw a straight line that is perpendicular to straight line ⓐ and passes through point B by using a ruler and a compass.

3 Hiroto used a ruler and a compass to draw a straight line that is parallel to straight line ⓐ and passes through point A as shown in the following diagram. Let's explain how he drew it.

What kind of figure becomes after connecting points in the order of A, B, D, C, and A?

Yui

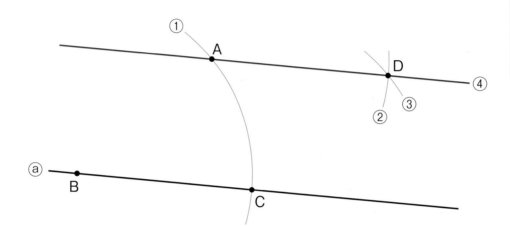

Way to see and think

Why can you draw a parallel straight line when you draw points B, C, and D?

4 In the following diagram, let's draw a straight line that is parallel to straight line ⓐ and passes through point A by using a ruler and a compass. Also, let's confirm that the straight line is parallel to straight line ⓐ.

A

Summary

By using the properties of a rhombus and parallelogram, perpendicular and parallel straight lines can be drawn only with a ruler and a compass.

In Junior High School, you will be able to draw more figures.

2 Daiki used a ruler and a compass to draw a straight line that divides an angle in half as shown in the following diagram. What kind of figures and properties did he use? Let's discuss.

Daiki's idea

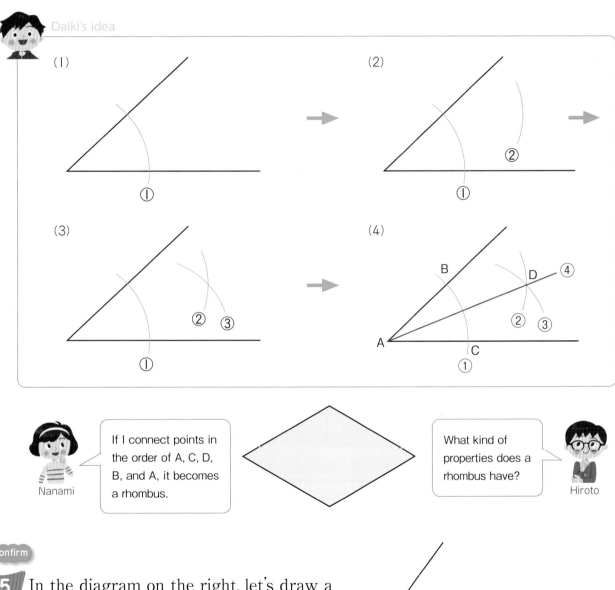

(I)

(2)

(3)

(4)

Nanami: If I connect points in the order of A, C, D, B, and A, it becomes a rhombus.

Hiroto: What kind of properties does a rhombus have?

5 In the diagram on the right, let's draw a straight line that divides the angle in half by using a ruler and a compass.

Can you represent with a graph?

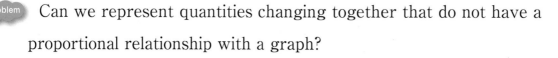

Problem · Can we represent quantities changing together that do not have a proportional relationship with a graph?

38

1 There is a container as shown on the right. When water is poured at a constant rate, what kind of graph will it become? Let's predict it.

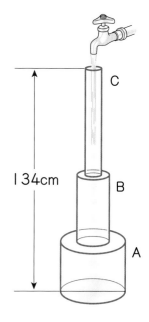

134cm

C
B
A

It doesn't seem to be one straight line.

Daiki

The way of increasing the water depth may change depending on the level A, B, and C.

Yui

Purpose Even when the way of increasing the water depth is changing, can we draw the graph?

Way to see and think

They think about different containers depending on the shape.

Let's think about it by separating the above container into containers A, B, and C as shown below.

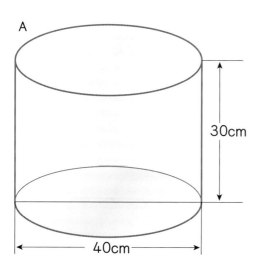

A

30cm

40cm

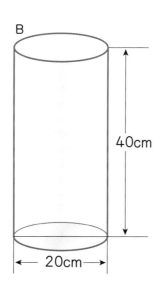

B

40cm

20cm

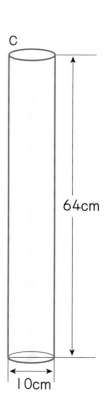

C

64cm

10cm

1 Water is poured into container A at a constant rate. Let's think about the following.

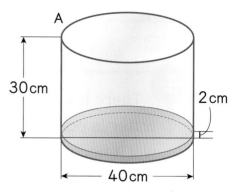

30 cm

2 cm

40 cm

① The table below shows the relationship between time and water depth. Let's complete this table.

Time and depth of poured water in container A

Time (minutes)	0	1	2	3	4	5											
Depth (cm)	0	2	4	6	8	10											

② Let's consider time as x minutes and depth as y cm, and represent the relationship between x and y in a math sentence.

> How many centimeters does the depth increase in 1 minute?

③ Water overflows from container A after several minutes. Let's represent the range of the values of x, by using the terms "greater than or equal to" and "less than or equal to," when water is poured into this container.

The values of x are greater than or equal to ☐ and less than or equal to ☐ .

④ Let's draw a graph which represents the relationship between the time and depth of the water that is poured into container A.

We can only draw the graph for the range of data in ③.

Nanami

Time and depth of poured water in container A

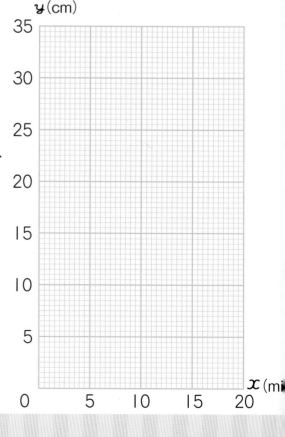

2 Water is poured into container B at the same constant rate as in problem . As for container B, how many centimeters does the depth increase in 1 minute?

B

40cm

20cm

Want to explain

① Hiroto solved the problem as shown below.

Let's explain how he solved the problem.

 Hiroto's idea

The depth increases **y** cm per minute in container B.

$$10 \times 10 \times 3.14 \times y = 20 \times 20 \times 3.14 \times 2$$
$$314 \times y = 2512$$
$$y = 8$$

We can write a math sentence if we can find which quantities are equal.

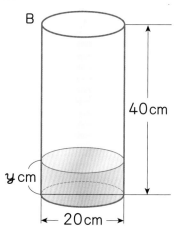

B

40 cm

y cm

20 cm

② In container B, the water also overflows after a certain period of time. Let's consider time as **x** minutes to think about the range of values of **x**, and complete the table below.

Also, let's represent the relationship between the time **x** minutes and the depth **y** cm in a math sentence.

Time and depth of poured water in container B

Time **x** (minutes)	0	1				
Depth **y** (cm)	0	8				

Time and depth of poured water in container B and C

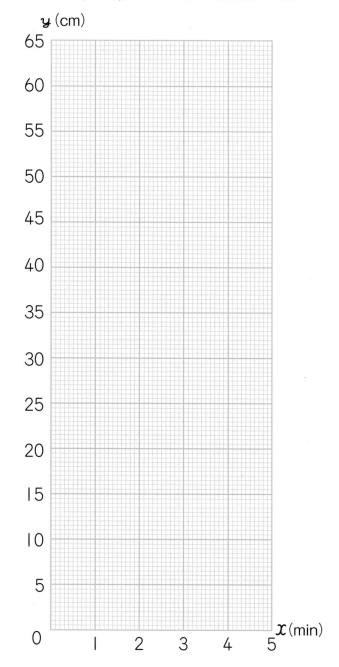

y (cm)

x (min)

Time and depth of poured water in container C

Time x (minutes)	0		
Depth y (cm)	0		

③ Let's draw a graph that represents the relationship between the time and depth of poured water in container B in the left diagram.

④ Water is poured into container C at the same constant rate as in problem ⚑. Let's write a math sentence and complete the table which represents the relationship between the time x minutes and the depth y cm. Also, let's draw a graph on the left.

C

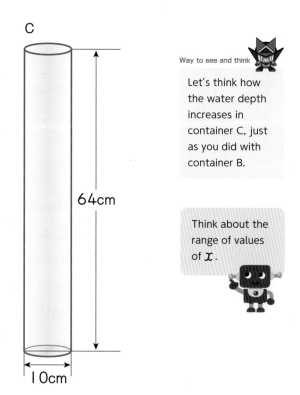

64cm

10cm

Way to see and think

Let's think how the water depth increases in container C, just as you did with container B.

Think about the range of values of x.

3 Let's draw a graph which represents the relationship between the time and depth of poured water in the original container composed of A, B, and C, when water is poured into this container at the same constant rate as in problem .

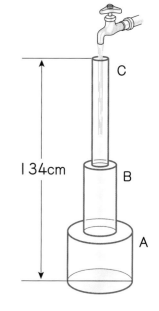

134cm

C

B

A

How should we draw the graph when container A is taken over by container B and container B is taken over by container C?

Yui

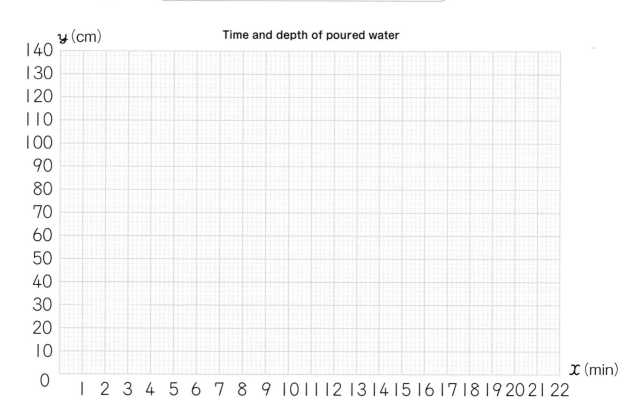

y (cm)   Time and depth of poured water

x (min)

As shown in the following diagram, when container A gets full, the drawing of graph for container B starts from there. When container B gets full, the drawing of graph for container C starts from there.

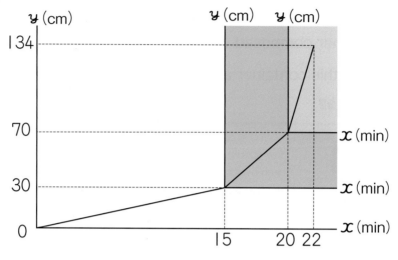

Summary

When the way of increasing the water depth is changing, you can draw a graph considering the range of time in which each container gets completely full.

In Junior High School, you will learn about not only straight line graphs but also curved graphs.

Want to deepen

2 When the order of containers A, B, and C is changed and water is poured at the same constant rate as in problem **1**, then the graph looks like the one shown below. Let's think how the containers were ordered. Also, what are the values of A~F?

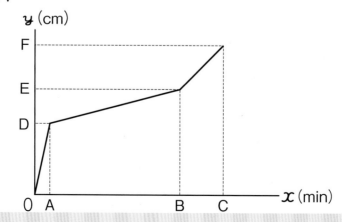

Way to see and think

Changing the order of containers changes the shape of the graph.

Can you stop exactly?

 Problem What should we do to find out which can stop closer to 10 seconds?

Want to solve

Plan of investigation 〔Plan〕

In the 10-second game, we want to investigate which can stop closer to 10 seconds, students of group 2 in 6th grade or the teachers?

Let's think about how to investigate.

First, let's play the 10-second game and take the records.

Daiki

We have to get help from the teachers.

Nanami

We have to get help from the teachers.

When asking, let them know our purpose.

Yui

You may anticipate what the result will be.

Hiroto

 Purpose By what kind of viewpoint should we investigate the collected data?

Data collection 〔Data〕

The following table shows the data of students of group 2 in 6th grade and teachers. Let's think about how to organize these data.

Data of group 2 in 6th grade (32 people) (seconds)

9.34	10.20	11.67	8.56	8.40	10.46	10.77	9.68
10.49	9.97	9.87	11.32	9.20	10.38	7.65	9.70
8.76	9.59	12.10	9.45	10.87	10.95	10.50	9.07
11.84	9.75	8.49	11.06	9.47	8.81	9.90	10.03

Data of teachers (20 people) (seconds)

9.26	10.03	9.44	10.19	9.38	9.80	9.98	9.75
10.74	9.93	10.86	9.88	9.96	10.10	8.95	9.33
10.62	11.16	10.34	11.25				

【Analysis】

Let's find the mean value of the data of group 2 in 6th grade and the teachers.

Let's summarize the data in a frequency table and represent it in a histogram.

Mean value of group 2 in 6th grade:

[ ] seconds

Mean value of teachers:

[ ] seconds

Data of group 2 in 6th grade

Class (seconds)			Number (people)
Greater than or equal to 7.5 ~ Less than 8.0			
8.0 ~ 8.5			
8.5 ~ 9.0			
9.0 ~ 9.5			
9.5 ~ 10.0			
10.0 ~ 10.5			
10.5 ~ 11.0			
11.0 ~ 11.5			
11.5 ~ 12.0			
12.0 ~ 12.5			
Total			

Data of teachers

Class (seconds)			Number (people)
Greater than or equal to 7.5 ~ Less than 8.0			
8.0 ~ 8.5			
8.5 ~ 9.0			
9.0 ~ 9.5			
9.5 ~ 10.0			
10.0 ~ 10.5			
10.5 ~ 11.0			
11.0 ~ 11.5			
11.5 ~ 12.0			
12.0 ~ 12.5			
Total			

Data of group 2 in 6th grade

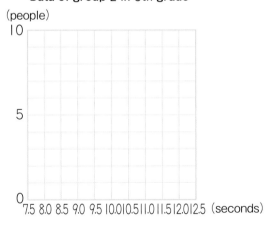

Data of teachers

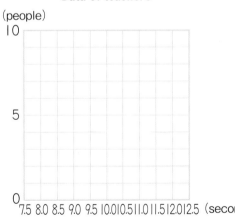

【Conclusion】

From the above results, which one is more likely to stop exactly at 10 seconds? Let's discuss.

【Problem】

Yui thought that group 2 in 6th grade was able to stop more accurate than the teachers because the number of children is larger in the class "Greater than or equal to 9.5 seconds to Less than 10.5 seconds." Is this idea correct? Let's discuss.

Daiki

There are 12 more people in group 2 in 6th grade than teachers. Is that a good thing?

Each total number is different. What about that?

Nanami

Analyzing 【Analysis】

Hiroto used a ratio to think as follows. Let's complete Hiroto's table.

 Hiroto's idea

Data of group 2 in 6th grade

Class (seconds)			Number (people)	Ratio
Greater than or equal to 7.5	~	Less than 8.0	1	
8.0	~	8.5	2	
8.5	~	9.0	3	0.09
9.0	~	9.5	5	
9.5	~	10.0	7	
10.0	~	10.5	5	
10.5	~	11.0	4	0.13
11.0	~	11.5	2	
11.5	~	12.0	2	0.06
12.0	~	12.5	1	
Total			32	1.00

Data of teachers

Class (seconds)			Number (people)	Ratio
Greater than or equal to 7.5	~	Less than 8.0	0	0.00
8.0	~	8.5	0	
8.5	~	9.0	1	0.05
9.0	~	9.5	4	
9.5	~	10.0	6	0.30
10.0	~	10.5	4	
10.5	~	11.0	3	
11.0	~	11.5	2	
11.5	~	12.0	0	0.00
12.0	~	12.5	0	
Total			20	

Findings 【Conclusion】

Based on Hiroto's table and the results so far, which one is more likely to stop exactly at 10 seconds?

Memo

Memo

Memo